Elementary Physics

Gases

BLACKBIRCH®
PRESS

THOMSON
———— ✱ ————
™
GALE

San Diego • Detroit • New York • San Francisco • Cleveland • New Haven, Conn. • Waterville, Maine • London • Munich

THOMSON

GALE

Photo Credits: **Art Explosion:** 3t, 9; **The Brown Reference Group plc:** 2, 7, 11, 16; **Corbis:** Brian Leng 12, Bruce Miller 10; **Image Bank:** Frans Lemmens 8, Tom Mareschal 14; **NASA:** 3b, 6, 18; **Pacific Northwest National Laboratory:** 1, 4.

Consultant: Don Franceschetti, Ph.D., Distinguished Service Professor, Departments of Physics and Chemistry, The University of Memphis, Memphis, Tennessee

For The Brown Reference Group plc
Text: Ben Morgan
Project Editor: Tim Harris
Picture Researcher: Helen Simm
Illustrations: Darren Awuah and Mark Walker
Designer: Alison Gardner
Design Manager: Jeni Child
Managing Editor: Bridget Giles
Production Director: Alastair Gourlay
Children's Publisher: Anne O'Daly
Editorial Director: Lindsey Lowe

LIBRARY OF CONGRESS CATALOGING-IN-PUBLICATION DATA

Morgan, Ben.
  Gases / by Ben Morgan.
      p. cm. — (Elementary physics)
Includes bibliographical references and index.
Summary: A simple introduction to what gases are, how they behave when heated or under pressure, and some other properties of gases.
  ISBN 1-41030-083-8 (hardback: alk. paper) — ISBN 1-41030-201-6 (softback: alk. paper)
  1. Gases—Juvenile literature. [1. Gases.] I. Title. II. Series: Morgan, Ben. Elementary physics.

QC161.2.M67 2003
530.4'3—dc21   2003002019

Printed and bound in Singapore
10 9 8 7 6 5 4 3 2 1

Contents

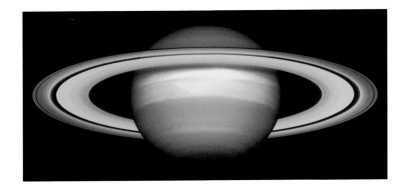

When you turn on a gas stove, an
invisible gas hisses out. When the gas
is lit, it burns with a blue flame.

What Are Gases?

Have you ever wondered why soda drinks are fizzy? It is because they contain bubbles of something called **gas**.

Everything is made of **matter**. There are three kinds of matter—**solids**, **liquids**, and gases. Things that you can touch and hold, like toys, are solids. They keep their own shape. Liquids are things that you can pour, like water. They take the shape of whatever you put them in. Gases are different. You cannot usually see or hold gases. But they are all around you, and they fill all the empty spaces.

Air

Balloons stretch and grow when you blow into them. They get bigger because they fill up with a **gas** called air.

Air is usually **invisible**, but you can see it if you know where to look. If you blow through a straw into a cup of water, you can make air appear as bubbles. You cannot normally hold air, but you can feel it when the wind blows against you. If you squeeze a balloon very gently, you can feel the air inside it push back.

From space, the air around Earth (left) looks like a blue layer.

a hot-air balloon

Hot Air

When air gets warmer, it rises.
Hot-air balloons rise because they
are full of warm air. To make a
hot-air balloon rise, the pilot turns
on a burner. The burner heats the
air inside the balloon. To come
down again, the pilot opens a
hole in the top of the balloon.
This lets the hot air out. The pilot
can make the balloon go up and
down, but he or she cannot steer
it. A hot-air balloon drifts wherever
the wind blows it.

Just as hot air rises, cold air sinks.
That is why basements are colder
than bedrooms. The coldest air in
a house sinks into the basement.
The warmest air rises upstairs.

When air
pressure is low,
it often rains.

Under Pressure

Air is very light, but it is not weightless. A roomful of air weighs about as much as a person. If you took all the air out of a room and put it into a bottle, you would not be able to pick up the bottle.

a barometer

The weight of all the air in the sky pushing down on Earth is called **air pressure**. Air pressure can be measured with a **barometer**. A barometer looks a bit like a clock. When the air pressure is high, the weather is often sunny. That is because the air in the sky is so heavy that it sinks. This keeps clouds from forming. When air pressure is low, clouds form and cause rainy weather.

Wet clothes dry fastest on a hot day.

Disappearing Water

Water is a **liquid**, but it can turn into a **gas** and disappear into the air. If you hang wet clothes out to dry, the water in them **evaporates**. That means the water turns into a gas called **water vapor**. The warmer the weather, the faster the water evaporates. That is why clothes dry quickly on a hot, sunny day.

Water does not disappear forever when it evaporates. If it cools down, it turns back into a liquid. On a cold day, you can see drops of water appear on the inside of windows. They form when water vapor in the air touches the cold glass.

Dandelion seeds are so light
they float on air.

Floating on Air

Most things fall straight to the ground when you drop them, but some things do not. Feathers, dandelion seeds, and **parachutes** can almost float on air. What keeps them from falling straight down?

When anything moves through air, the air pushes back a little. If you run very fast, you can feel the push of the air as wind on your face. Objects that are very light catch a lot of air, so they get pushed the most. As they fall, the air pushes back at them and keeps them afloat. Parachutes catch so much air that they fall very slowly. People use them to float to the ground from high up.

15

Hot pizza
smells good
because the heat
makes gases
escape from it.

Gas Detector

You cannot usually see **gases**, but you can often smell them. Hot food often has a very strong smell. The heat of the food makes invisible gases escape into the air. The gases travel through the air and enter your nose.

As you chew something, gases from the food travel from the back of your mouth and into your nose. These smells make the food taste better. Pinch your nose closed the next time you eat. You will find that the food is not as tasty.

Jupiter is the biggest planet, and
Saturn (below) is the lightest.

Great Red Spot ————————————

Gas Giants

We live on Earth, which is one of nine **planets** that circle the Sun. Our planet is made of rock. The biggest planets are made of nothing but **gas**. Jupiter and Saturn are both made of gas. **Astronomers** call them gas giants because they are so huge. Jupiter is more than a thousand times bigger than Earth. Saturn is almost as big as Jupiter, but much lighter. If you could put Saturn in a bucket of water, it would float.

The gas in Jupiter flows around its surface in giant bands. Sometimes the bands mix together and form huge swirling storms. Jupiter's Great Red Spot formed this way.

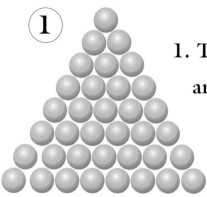

1. The molecules in a solid
   are packed together like
   bricks in a wall.

2. The molecules in
a liquid can move
around separately,
but they always stay
close together.

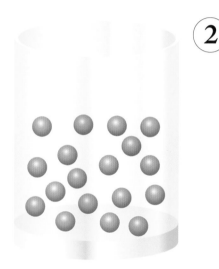

3. The molecules
in a gas fly off
on their own and
spread out.

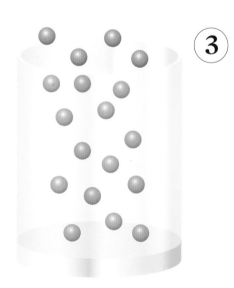

Atoms and Molecules

Everything in the world is made of very tiny particles called **atoms**. Usually the atoms form small groups called **molecules**. Molecules are so small that you cannot see them, even with a microscope. **Solids**, **liquids**, and **gases** are different from each other because of the way their molecules are arranged. In a solid, all the molecules are joined together strongly and held in place like bricks in a wall. In a liquid, the molecules can move around separately, but they always stay close together. In a gas, the molecules can fly off on their own and spread out.

Make a Spinning Snake

A simple toy can show that hot air rises. You need a large piece of paper, a pencil, some scissors, and some thread. First, draw a spiral on the paper like the one here. It is easier if you start in the middle. Next, cut along the line to make a snake. You can color it in to look like a real snake. Then hang the snake above a heater with the thread (ask an adult to help you). When the heater is hot, warm air rises from its top. The rising air pushes the snake and makes it spin.

warm air

22

Glossary

astronomer a person who studies outer space.

atom the smallest particle of a liquid, solid, or gas. Atoms sometimes group together to form molecules.

barometer an instrument that measures the pressure of air.

evaporate to turn from a liquid into a gas.

gas a substance that will spread to fill any space that contains it.

invisible something that is present but cannot be seen.

liquid a substance that can be poured. It will take the shape of the container that contains it.

matter anything, whether liquid, solid, or gas. All matter is made of atoms and molecules.

molecule a group of atoms.

microscope an instrument that makes enlarged images of small objects.

parachute a structure that slows an object as it falls through air.

planet large body that revolves around the Sun.

solid a hard substance that keeps its shape.

water vapor the gas that forms when water evaporates.

Look Further

To find out more experiments you can carry out with gases, read *101 Great Science Experiments* by Neil Ardley (DK Publishing).

You can also find out more about the different kinds of matter from the internet at this website: www.chem4kids.com/

Index